AIgents

Your Guide to AI Agents
in the Workplace

Tony Eggleston

[Amazon Best-Selling Author of *"How YOU Can Use ChatGPT"*]

Copyright © 2024 by **Tony Eggleston**

All Rights Reserved

No part of this publication may be reproduced, distributed, or transmitted in any form or by any means, including photocopying, recording, or other electronic or mechanical methods, without the prior written permission of the publisher, except in the case of brief quotations embodied in critical reviews and certain other noncommercial uses permitted by copyright law.

AIgents: Your Guide to AI Agents in the Workplace

The business world is changing faster than ever before. Every day, I work with companies struggling to keep up with increasing demands, growing competition, and the relentless pace of technological change. But I'm also seeing something exciting: businesses that are thriving by embracing a new way of working - one that combines human intelligence with AI in ways that seemed impossible just a few years ago.

This book is your guide to that new way of working. As the author of the Amazon best-selling book *"How YOU Can Use ChatGPT,"* I've helped thousands of people understand and implement AI technology. Now, with AIgents, I'm going to show you how to transform your entire business operation.

This isn't just another book about AI. It's a practical guide to implementing AI agents - or as we call them, AIgents - in ways that create real value for your business. You'll learn exactly how to use this technology to save time, reduce costs, and improve results across every aspect of your operation.

More importantly, you'll learn how to do it right. Through real examples and detailed guidance, I'll show you how to implement AIgents in ways that enhance your team's capabilities rather than replace them. You'll see how businesses like yours are using AIgents to save up to 38% of their employees' time while improving service quality and operational efficiency.

Whether you're new to AI or already using basic AI tools, this book will show you the path forward. Let's begin.

To those who understand that the future of business lies in the harmonious collaboration between human intelligence and artificial intelligence.

TABLE OF CONTENTS

1

Why AIgents? The Game-Changer Your Business Needs

Let me tell you a story. Last month, I was working with a business owner who was skeptical about AI. "Tony," she said, "I've tried automation before. It was clunky and created more problems than it solved." I hear this a lot. However, after implementing AIgents in her customer service department, she called me back within two weeks. Her team's productivity had skyrocketed, customer satisfaction was up, and her employees were happier because they could focus on the interesting parts of their jobs instead of repetitive tasks.

This is happening across every industry I work with. Remember when email first became common in offices? Some people thought it was just a fad. Others couldn't imagine how it would change the way we work. We're at another moment like that right now. AIgents are about to transform the workplace just as dramatically as email did, and I want to make sure you're ahead of the curve.

The business world is moving faster than ever. If you're like most business owners and managers I talk to, you're feeling the pressure to do more with less. Your customers expect instant responses at any hour. Your employees want meaningful work, not endless data entry. And you need to stay competitive without breaking the bank.

Here's what makes AIgents different from any other solution you've tried before. When I wrote my book about ChatGPT, I showed people how AI could help with individual tasks. AIgents take this to a whole new level. They're not just handling one task at a time - they're managing entire workflows. They're learning your business processes, understanding your customers' needs, and continuously improving their performance.

Let me show you what this looks like in practice. One of our clients, a healthcare clinic, was struggling with appointment scheduling. Their front desk staff was spending hours each day on the phone, playing phone tag with patients, and manually entering appointments. We implemented an AIgent that could handle all of this automatically. It understands natural language, knows the doctors' schedules, manages cancellations and reschedules, and even sends reminders to patients. The result? The front desk staff now spends their time actually helping patients in the office, and appointment no-shows have dropped by 20%.

But here's what really gets me excited - the time savings. Through our work with businesses across different industries, we've found that AIgents can save up to 38% of your employees' time. Think about that. For every 100 hours of work, you're getting almost 40 hours back. What could your team accomplish with that extra time?

I see the impact of this every day. Marketing teams finally have time to work on campaigns they've been putting off. Sales teams can focus on building relationships instead of updating CRM records. Customer service representatives can handle complex issues properly instead of rushing through calls. This isn't just about saving time - it's about enabling your team to do their best work.

Now, you might be thinking about the cost. Trust me, I understand. Any business investment needs to make financial sense. But here's what makes AIgents different: they actually help reduce costs while allowing your business to scale. Unlike hiring more staff, where costs increase linearly with workload, AIgents can handle growing volume without proportional cost increases. Once they're set up, they can process ten customer inquiries or ten thousand with the same efficiency.

Let me be very clear about something though - AIgents aren't about replacing people. In my experience helping thousands of people understand AI through my book and consulting work, I've learned that AI works best when it enhances human capabilities rather than trying to replace them. AIgents take over the routine work that drains your team's time and energy, letting them focus on tasks that require human creativity, empathy, and strategic thinking.

At AIgent, LLC, we've developed a unique approach to implementing these AI assistants. Just like I broke down ChatGPT into simple, actionable steps in my best-selling book, we make implementing AIgents straightforward and practical. We work with you to understand your specific needs, customize the AIgents for your business, and ensure they integrate smoothly with your existing operations.

Think about your business right now. What could your team accomplish if they had 38% more time? How would your customer service improve if you could provide instant, accurate responses at any hour? What opportunities could you pursue if routine tasks didn't consume so much of your day? These are the questions that get me excited about AIgents, and I hope they get you thinking too.

The future of work is already here, and it's more exciting than intimidating. Throughout this book, I'll show you exactly how AIgents can transform your business, using real examples and practical strategies you can implement right away. Just like I did with ChatGPT, I'm going to break everything down into clear, actionable steps. Ready to learn more? Let's dive deeper in the next chapter.

2

Inside AIgent, LLC - Why Experience Matters

When my book *"How YOU Can Use ChatGPT"* hit the Amazon bestseller list, I got flooded with messages from business owners. They'd read my book and see what AI can do, but they wanted more. They needed someone who could help them implement AI in their businesses the right way. That's exactly why I created AIgent, LLC.

I've got to tell you something important about AI implementation - there's a right way and a wrong way to do it. I've seen too many businesses waste time and money on AI solutions that look impressive in demos but fall apart in the real world. You know what I'm talking about if you've ever tried one of those chatbots that just keeps repeating the same useless responses. That's not what we do at AIgent, LLC.

Here's what makes us different: I'm a certified MindStudio AI Expert - that's their highest level of certification and we are a Mindstudio Partner. But more importantly, I know how to translate complex AI technology into real business results. The same way I broke down ChatGPT into simple, actionable steps in my book, I help businesses implement AIgents in ways that make immediate, practical sense.

Let me tell you about a client who came to us recently. They'd already spent a fortune on an AI system from one of the big tech companies. Fancy interface, lots of features - and their employees hated it. Why? Because it wasn't built for their specific needs. It was like trying to force a square peg into a round hole. Within three months of working with us, they had AIgents that their team actually enjoyed using because we took the time to understand their business first.

Our partnership with MindStudio is a game-changer, but let me tell you why that matters to your business. MindStudio isn't just another AI platform - they're the leader in creating specialized AI Workers. Think of it like this: if most AI solutions are like off-the-rack suits, MindStudio lets us create custom-tailored solutions that fit your business perfectly. And because I'm certified at their highest level, we can do things with their technology that other companies simply can't.

Here's a real example of how this works. A financial services company came to us drowning in compliance checks. Their team was spending hours reviewing documents, and they were terrified of missing something important. We created an AIgent specifically for their compliance process. Not a generic AI - an AIgent that learned their exact requirements, could review documents in seconds and knew exactly what to flag for human attention. Within three months, they reduced their review time by 35% and actually caught compliance issues they might have missed before.

But here's what I'm most proud of - we stick around. Remember when I wrote my ChatGPT book? I didn't just explain how to use it; I showed people how to get better with it over time. We take the same approach with AIgents. This isn't a *set it and forget it* solution. We monitor performance, make adjustments, and ensure your AIgents keep delivering value as your business evolves.

You know what really drives me crazy? Tech companies that hide problems behind jargon and buzzwords. That's not our style. Just like in my book, I believe in straight talk. If something isn't working, we'll tell you and fix it. If there's a better way to do something, we'll let you know. This isn't just about integrity - it's about getting you real results.

Let me tell you something that happened recently that really shows how we work. A retail client came to us wanting to automate their customer service. Pretty standard request, right? But when we dug deeper, we discovered their real problem wasn't just handling customer inquiries - it was maintaining consistency across their rapidly growing business. They had multiple locations, each doing things slightly differently, and it was causing chaos.

We could have just given them a basic chatbot and called it a day. Instead, we created a network of AIgents that not only handled customer service but also helped standardize operations across all their locations. The AIgents learned the best practices from each store and helped create a unified approach. Customer satisfaction went up, employee frustration went down, and they saw operational improvements they hadn't even expected.

This gets at something fundamental about our vision for AIgents. When I wrote about ChatGPT, I showed how AI could be a powerful personal assistant. AIgents take this to the organizational level. They become like ideal team members - always available, consistently reliable, and constantly learning. But they're not replacing your human employees; they're empowering them to do more meaningful work.

I remember working with an accounting firm where the junior accountants were initially terrified of AIgents. They thought they were going to be automated out of a job. But once they saw how AIgents could handle the tedious parts of their work - data entry, basic reconciliations, standard reports - they became our biggest supporters. Why? Because they could finally focus on the analytical work they went to school for. Their jobs became more engaging, not less secure.

Looking ahead, I see AIgents becoming as essential to business as email and smartphones. Not because they're trendy, but because they solve real problems in ways that make economic sense. They help businesses scale without proportionally increasing costs. They improve consistency while reducing errors. They enable levels of service that would be impossible to maintain with human staff alone.

But here's what really matters: they do all this while making work more satisfying for your human team. That's the future we're building at AIgent, LLC. A future where technology and human talent work together seamlessly, each focused on what they do best.

In the next chapter, we'll dive into the technology that makes AIgents possible. Don't worry - I'll break it down the same way I did with ChatGPT in my book. You don't need to be a tech expert to understand how AIgents can transform your business. You just need to see the possibilities.

3

The Technology Behind AIgents - What You Actually Need to Know

et me explain complex technology in a way that actually makes sense. You don't need to understand every technical detail to use AIgents effectively - just like you don't need to know how a car engine works to be a great driver. However, understanding the basics will help you see the possibilities for your business.

Let me start with something I hear all the time: *"Aren't AIgents just fancy chatbots?"* No way. That's like saying a smartphone is just a fancy calculator. Sure, AIgents can chat, but that's just scratching the surface. The technology behind AIgents is what gives them their real power, and it comes down to three main components: machine learning, natural language processing, and advanced automation. Don't worry if those terms sound intimidating - I'm going to break them down in plain English.

First up, machine learning. This is what makes AIgents different from traditional software. Regular software follows rules - if this happens, do that. AIgents learn from experience, just like humans do. Let me give you an example from a client in the healthcare industry. When they first implemented their AIgent for patient scheduling, it followed basic rules about available time slots. But within weeks, it had learned patterns - which patients were more likely to show up on time, which appointment types typically ran long, and even which days tended to have more cancellations. It started making smarter scheduling decisions based on this learning, and their no-show rate dropped dramatically.

The second piece is natural language processing, or NLP. This is what lets AIgents understand and communicate like humans. They don't just understand words - they get context, tone, and intent. We had a retail client whose AIgent learned to recognize when a customer was frustrated versus just asking for information, and it would adjust its responses accordingly. It's like having a customer service rep who always knows exactly how to respond.

Then there's advanced automation. But forget what you know about traditional automation - this is different. Traditional automation is like a train on tracks - it can only go where the tracks lead. AIgents are more like skilled drivers who can navigate any road and adjust to changing conditions. One of our manufacturing clients used to have a rigid automation system for their inventory management. If something unexpected happened, the whole system would grind to a halt. Their AIgent can handle surprises - supply chain delays, sudden demand spikes, even equipment maintenance needs - and adjust the workflow automatically.

Now, all this sophisticated technology runs on MindStudio's platform. As a certified MindStudio AI Expert, I can tell you there's a reason we chose to partner with them. MindStudio isn't just another AI platform - they're the leaders in creating specialized AI Workers. While other platforms offer one-size-fits-all solutions,

MindStudio lets us create AIgents that are truly customized for your business.

Let me give you a real example of how all this technology comes together. We worked with a financial services company that was struggling with document processing. They had thousands of documents coming in daily - loan applications, tax forms, identity verification - and their team was drowning in paperwork. We created an AIgent that could not only read these documents but understand them contextually.

This AIgent learned their specific requirements and compliance standards. It could spot patterns that might indicate fraud. It knew when something needed human review and when it could process automatically. But here's the really impressive part - it kept getting better. When it encountered new types of documents or situations, it learned from them. If a human reviewer made a correction, it remembered that for next time. Within months, it was handling 70% of their document processing without any human intervention, and doing it more accurately than their previous manual process.

Let's talk about something crucial - security. Security is a top priority because AIgents handle business data. We work with each client to understand their specific security needs and regulatory requirements and then implement appropriate protections. This includes access controls, secure data handling procedures, and monitoring systems. Each business has unique security requirements, and we help them make smart decisions about how their AIgents handle sensitive data.

Here's something else that surprises people about AIgents - they can work together as a team. While one AIgent might handle customer service, another manages inventory, and a third processes orders, they can all communicate and coordinate with each other. We helped a retail chain set up a network of AIgents that worked this way. When a customer asked about a product, the customer service AIgent could instantly check with the inventory AIgent about stock levels and the shipping AIgent about delivery times. It was like having a perfectly coordinated team working 24/7.

The real power of this technology shows in its ability to make smart decisions. AIgents can analyze vast amounts of data, identify patterns, and make recommendations that humans might miss. Let me give you a real example. We worked with a company that thought they had a staffing problem in their call center. Their AIgent analyzed their call patterns and discovered something surprising - their peak call times weren't when they thought they were. By adjusting their schedules based on the AIgent's analysis, they improved their response times without hiring a single new person.

The technology behind AIgents is constantly evolving, and that's another benefit of working with us. Our partnership with MindStudio means your AIgents can take advantage of new capabilities as they're developed. It's like getting a software update that makes your phone more capable, except these updates can transform how your business operates.

Here's what really matters though - all this sophisticated technology serves one purpose: making your business run better. You don't need to understand every technical detail to see results. What matters is how AIgents help your business grow, save your team time, and improve your customer service.

I worked with a manufacturing company recently that illustrates this perfectly. They implemented AIgents to handle their supply chain management. The AIgents could track inventory levels, predict demand patterns, and automatically adjust orders based on multiple factors - something that would have taken their team hours of analysis to do manually. But the real breakthrough came when they discovered the AIgents could spot potential supply chain disruptions before they became problems. It started flagging unusual patterns in supplier delivery times and suggesting inventory adjustments to compensate.

That's the kind of intelligence that transforms a business. It's not just about automating tasks - it's about having AI assistants that can think ahead and help prevent problems before they occur.

In the next chapter, we'll look at how AIgents work alongside your human team. Because while the technology is powerful, it's the combination of human intelligence and AI capabilities that really transforms a business. I'll show you exactly how to create that perfect balance where both humans and AIgents can do what they do best.

4

Redefining Roles – How Humans and AIgents Work Together

Let's address the elephant in the room right away. Whenever I talk about AIgents, someone inevitably asks, "Are these going to replace my employees?" The short answer is no. The long answer is that AIgents actually make human jobs more interesting and valuable. Let me show you why.

Think about your most talented employees. I bet they spend a good chunk of their time doing routine tasks that don't really use their skills. Maybe your best salesperson is updating spreadsheets instead of talking to clients. Or your experienced customer service reps are answering the same basic questions over and over instead of handling complex issues where their expertise really matters.

This is exactly where AIgents come in. They're not about replacing people - they're about letting people do what they do best. Let me give you a real example from a law firm we worked with recently. Their paralegals were spending hours every day doing document searches and basic contract reviews. It was mind-numbing work that had to be done, but it wasn't making good use of their training and expertise.

We implemented AIgents to handle the initial document review and search process. The AIgents could scan thousands of documents in minutes, flag relevant sections, and identify potential issues. Now, here's what's interesting - instead of replacing the paralegals, this made their work more valuable. They could focus on analyzing the findings, spotting nuances that required human judgment, and providing better insights to the attorneys.

The result? The firm could handle more cases, the paralegals were happier because they were doing more engaging work, and the quality of their legal support actually improved. That's what I mean when I say AIgents enhance human work rather than replace it.

Let me break down how this human-AIgent collaboration typically works. Think of AIgents as highly capable team members who excel at certain types of tasks:

Processing large amounts of data quickly Handling routine, repetitive tasks consistently Working 24/7 without getting tired Learning and applying patterns Managing multiple workflows simultaneously

Humans, on the other hand, are irreplaceable when it comes to: Strategic thinking and planning Creative problem-solving Building relationships Making nuanced judgments Handling complex emotional situations

When you pair these strengths together, something remarkable happens. I saw this with a retail client who implemented AIgents in their customer service department. The AIgents handled routine inquiries - tracking orders, answering common questions, processing simple returns. This freed up the human agents to focus on complex customer issues that required empathy and problem-solving skills.

But here's what really surprised them - their customer satisfaction scores went up across the board. Why?

Because customers got faster responses to simple questions from the AIgents, and when they needed human help, they got the full attention of agents who weren't rushed or bogged down with routine tasks.

The transition to working with AIgents isn't always smooth sailing though. I'll be straight with you about that. People can be skeptical or nervous at first. That's why we put so much emphasis on proper training and integration. We show teams exactly how AIgents will make their jobs easier, not obsolete.

One of our manufacturing clients had initial resistance from their inventory management team. These folks had been doing their jobs for years and were worried about being replaced by AI. So we took a different approach. Instead of just implementing the AIgents, we involved the team in the process. We showed them how AIgents could handle the tedious parts of their job - like data entry and routine ordering - while they focused on strategic inventory decisions and supplier relationships.

Within a month, these same team members were coming up with new ways to use the AIgents. They went from being skeptical to being innovators, finding creative ways to use AI to improve their work. One team member told me, "This isn't what I expected. It's like having a super-powered assistant who handles all the stuff I hate doing."

This brings me to an important point about leadership. Successfully integrating AIgents isn't just about the technology - it's about change management. As a leader, you need to help your team see AIgents as tools for empowerment rather than replacement. The best way to do this is through clear communication and early wins.

Start with a specific area where AIgents can clearly improve things for your team. Maybe it's taking over nighttime customer service so your team doesn't have to work late shifts. Or handling the month-end report generation that everyone dreads. Show your team concrete examples of how AIgents make their work lives better.

I've also found that involving your team in the AIgent implementation process yields better results. Your employees know their jobs better than anyone. They can identify the tasks that are ripe for automation and help ensure the AIgents are configured in ways that truly support their work.

Let me share a success story that brings all this together. We worked with a healthcare provider who wanted to improve their patient scheduling system. Instead of just implementing an AIgent system, we worked with their scheduling staff to understand their biggest challenges. They told us about the complexities of matching patients with the right specialists, managing cancellations, and dealing with insurance verifications.

We designed their AIgents to handle these specific pain points. The AIgents took over tasks like insurance verification, appointment reminders, and basic scheduling. But the human team remained essential for handling complex cases, building relationships with referring doctors and making judgment calls about urgent cases.

The result was a hybrid system that worked better than either humans or AI alone could achieve. The scheduling staff became more like patient care coordinators, focusing on the human aspects of healthcare while letting the AIgents handle the administrative burden.

Looking to the future, this kind of human-AIgent collaboration will become the norm. Businesses that get it right will have a significant advantage. Their employees will be more engaged because they're doing meaningful work. Their operations will be more efficient because routine tasks are handled automatically. And their customers will be happier because they get the best of both worlds - efficient service for simple needs and human attention when it matters most.

In the next chapter, we'll dive into how AIgents transform customer experience. You'll see exactly how this human-AIgent collaboration creates service levels that would be impossible to achieve any other way.

5

Transforming Customer Experience with AIgents

Let me tell you about a call I got from a business owner last month. His company was losing customers because they couldn't handle support requests fast enough. *"My team is working overtime,"* he said, *"but we still can't keep up. People are waiting hours for basic answers."* Sound familiar?

Three weeks after implementing AIgents, he called back excited. Their response time had dropped from hours to seconds. Customer satisfaction was up, and his support team was finally able to focus on complex issues instead of answering the same questions over and over. That's the power of AIgents in customer service - but let me show you exactly how it works.

First, let's talk about what customers expect today. They want answers immediately, not "during business hours." They want consistent service whether they're your first customer of the day or the last. And they want personalized attention when they have real problems. That's a tall order for any human team to handle, no matter how talented they are.

Here's where AIgents change the game. Imagine having customer service representatives who:

- **Never sleep**
- **Never get tired or frustrated**
- **Know your entire product catalog perfectly**
- **Can handle thousands of conversations simultaneously**
- **Learn and improve from every interaction**
- **Speak multiple languages fluently**

Sounds impossible? That's exactly what AIgents do. But here's the key - they don't replace your human team. They transform what your human team can accomplish.

Let me share a real example from a software company we worked with. Their support team was overwhelmed with basic questions about installation, password resets, and common features. The interesting part? When we analyzed their support tickets, we found that 67% of questions were variations of the same 20 issues.

We implemented AIgents to handle these common scenarios. The results were immediate. Response times dropped from 45 minutes to under 30 seconds for basic issues. But here's what really matters - customer satisfaction didn't just improve for the simple questions the AIgents handled. It improved across the board.

Why? Because when customers had complex problems that needed human attention, they got it faster and better. The support team wasn't burned out from answering the same basic questions all day. They had time to really dig into difficult issues and provide thoughtful solutions.

But great customer service isn't just about speed. It's about consistency and personalization. This is where AIgents really shine. Let me show you how.

Consistency first. One of our retail clients had a problem with different customers getting different answers to the same questions. It depended on which agent they talked to and what that agent remembered from training. AIgents solved this instantly. Every customer gets the same accurate information, every time. But - and this is crucial - the AIgents know when to get a human involved if something falls outside their knowledge base.

Now let's talk about personalization. AIgents remember every interaction with every customer. A banking client of ours implemented AIgents that could instantly pull up a customer's entire history and preferences. If someone called about a transaction, the AIgent already knew their account preferences, previous issues, and typical banking patterns. It could spot potential problems before the customer even mentioned them.

Here's something else that surprises people - AIgents can handle multiple channels seamlessly. Phone, email, chat, social media - they manage them all simultaneously while maintaining consistency. One of our retail clients used to have different teams for each channel. Now their AIgents handle basic inquiries across all channels, and customers get the same high-quality service no matter how they reach out.

But what about complex conversations? This is where our voice AIgents come in. These aren't your typical automated phone systems that make you want to throw your phone across the room. Our voice AIgents understand natural conversation, pick up on tone and context, and know when to transfer to a human agent.

A healthcare provider we work with uses voice AIgents for appointment scheduling. These AIgents don't just book appointments - they understand priority levels, manage cancellations, send reminders, and can even recognize when someone sounds distressed and needs immediate human assistance.

Let me tell you about something really interesting we discovered. When we first implement AIgents, companies often expect customer satisfaction to go down for AIgent-handled interactions compared to human agents. But we consistently see the opposite. Why? Because AIgents are:

- Always polite (even with rude customers)
- Never in a rush
- Consistently accurate
- Available instantly

- Able to handle multiple requests simultaneously

But there's one thing that makes an even bigger difference - AIgents get better over time. They learn from every interaction. If they don't handle something well, they adapt. If customers frequently ask about a new issue, they learn about it. It's like having customer service agents who instantly share knowledge and never forget what they learn.

The real magic happens when you combine AIgents with human agents effectively. We helped a travel company set this up perfectly. Their AIgents handle all the routine stuff - checking flight status, explaining baggage policies, processing simple refunds. But they're also trained to recognize when a customer needs human help - like when someone's stuck at an airport with a complex problem that needs creative problem-solving.

The system is so smooth that customers often don't even realize they're moving between AIgent and human support. It just feels like great service.

Here's what this means for your business: you can provide better customer service than ever before, at a scale that was impossible with human agents alone. Your customers get instant answers to simple questions and more attentive human service when they really need it. Your support team gets to focus on interesting challenges instead of repetitive tasks. And you get consistent, high-quality customer service 24/7 without the cost of staffing multiple shifts.

In the next chapter, we'll look at how AIgents handle specific tasks and workflows. You'll see exactly how they can transform your operations while making work more enjoyable for your team.

6

AIgents and Task Automation - Making Work Flow Better

et me paint a picture of something I see all the time. A business is running well, growing steadily, but there's a problem. Everyone's busy - really busy. Not with the strategic work that grows the business, but with endless routine tasks that have to get done. Data entry. Report generation. Invoice processing. Email responses. The stuff nobody loves doing but everybody has to deal with.

That's exactly where AIgents excel. But I'm not talking about simple automation that breaks the moment something unexpected happens. I'm talking about intelligent automation that can think, adapt, and handle complex workflows.

Let me tell you about a manufacturing company I worked with recently. Their purchasing team was spending hours every day just processing routine orders. Check inventory levels, create purchase orders, send them to suppliers, update the system, notify relevant departments. Basic stuff, but it had to be done right. They figured it was too complex for automation because there were too many variables to consider.

We set up an AIgent to handle their purchasing workflow. Within a month, it was processing 80% of their routine orders automatically. But here's the impressive part - it didn't just follow rules. It learned. It started recognizing patterns in supplier response times. It adjusted order quantities based on seasonal demands. It even started flagging potential supply chain issues before they became problems.

The purchasing team? They finally had time to negotiate better contracts, find new suppliers, and work on improving their supply chain. Their jobs became more interesting and more valuable to the company.

Here's another example that might hit close to home. A financial services firm was drowning in document processing. Every day, thousands of documents needed to be reviewed, categorized, data extracted, and entered into their system. They had a team of skilled people spending their days basically copying and pasting information.

We implemented AIgents to handle the document workflow. These AIgents could read documents, understand their content, extract relevant information, and enter it into the right systems. They could recognize different document types, handle variations in format, and even flag unusual items for human review.

But the real breakthrough came when the AIgents started identifying patterns in the documents that humans had missed. They spotted trends in customer requests, identified common error patterns, and even suggested improvements to their document templates. The team went from data entry clerks to data analysts, using their expertise to improve processes instead of just feeding information into systems.

Let's talk about something that every business deals with - scheduling and calendar management. Sounds simple, right? But anyone who's tried to coordinate meetings with multiple people knows it can eat up hours of time. One of our clients, a consulting firm, calculated that their team was spending an average of 5 hours per person per week just dealing with scheduling.

Their AIgent now handles all of that. It coordinates schedules, finds available times, sends invites, manages changes and cancellations, and even learns people's preferences over time. If someone usually needs prep time before client meetings, the AIgent builds that into their schedule. If certain team members prefer morning meetings, it takes that into account when suggesting times.

But here's where it gets really interesting. The AIgent started identifying patterns that helped the whole company work better. It noticed that certain types of meetings consistently ran over their scheduled time and started automatically adjusting calendar blocks for those. It recognized that specific team combinations led to more productive meetings and suggested optimal groupings.

One of the most powerful features of AIgents is their ability to handle complex workflows that span multiple departments. Take invoice processing, for example. A typical invoice might need to be reviewed by accounts payable, checked against purchase orders, approved by department heads, and entered into multiple systems. It's the kind of process that often gets bottlenecked because it requires coordination between different people and systems.

We helped a construction company set up AIgents to manage their entire invoice workflow. The AIgents could read invoices, match them to purchase orders, route them for appropriate approvals, flag discrepancies, and even learn which kinds of issues needed human attention versus what could be processed automatically.

The system got so good at identifying patterns that it started predicting potential issues before they happened. If an invoice seemed unusual compared to historical patterns, it would flag it for review. If a vendor's prices started creeping up, it would notify purchasing. The accounting team went from chasing papers to focusing on financial strategy.

Voice AIgents add another dimension to workflow automation. These aren't just phone systems - they're intelligent assistants that can handle complex conversations and tasks. A healthcare provider we work with uses voice AIgents for appointment management. These AIgents don't just schedule appointments - they understand medical terminology, recognize urgency in a patient's voice, and know when to transfer to a human immediately.

The key to successful task automation with AIgents is understanding that they're not just about replacing manual work - they're about transforming how work gets done. They learn and improve over time. They spot patterns and suggest improvements. They coordinate with each other and with human team members seamlessly.

Remember that manufacturing company I mentioned at the start? Six months after implementing AIgents, they discovered something fascinating. Not only were routine tasks being handled more efficiently, but their entire operation was working better. The AIgents had learned enough about their business to start suggesting process improvements nobody had thought of.

That's the real power of AIgents in workflow automation. They don't just take over tasks - they help your whole organization work smarter. In the next chapter, we'll look at some specific case studies that show exactly how different businesses have used AIgents to transform their operations.

7

Case Studies
- AIgents in Action

Theory is great, but nothing beats seeing real results. In this chapter, I want to show you exactly how different businesses have transformed their operations with AIgents. These aren't hypothetical scenarios - these are real companies solving real problems.

Let's start with one of my favorite success stories. A mid-sized insurance company was struggling with claims processing. Their team was working overtime just to keep up with basic claims, leaving complex cases waiting too long for attention. They were losing customers and burning out their staff.

Here's what we did. We implemented AIgents to handle the initial claims review process. The AIgents could read claim submissions, verify policy coverage, check for common issues, and either process straightforward claims automatically or route complex ones to human adjusters.

The results? Claims processing time dropped from an average of 5 days to less than 24 hours for routine claims. But here's what really matters - customer satisfaction jumped 40%, and employee turnover in their claims department dropped to nearly zero. Why? Because the adjusters could finally focus on complex cases where their expertise really mattered.

Now let me tell you about a retail chain that had a different problem. They were growing fast, opening new stores across the country. However, maintaining consistent customer service across locations was becoming impossible. Every store had slightly different ways of doing things, and training new staff was taking too long.

We created a network of AIgents that became their central knowledge base and service system. These AIgents handled customer inquiries across all channels - phone, email, chat, social media. They maintained perfect consistency in answers and could instantly adapt when policies changed.

The fascinating part? The AIgents actually helped standardize operations across all locations. They learned the best practices from each store and helped create a unified approach. Within three months, customer satisfaction scores were virtually identical across all locations - something they'd never achieved before.

Here's a case that really shows the power of AIgents in handling complex workflows. A commercial real estate company was struggling with lease management. Every lease agreement involved multiple departments - legal, finance, property management, maintenance. Documents were getting lost in email chains, approvals were delayed, and important renewal dates were being missed.

Their AIgent system transformed this entire process. It could read and understand lease documents, track key dates, coordinate approvals, and ensure nothing fell through the cracks. But it did more than that - it started identifying patterns in lease negotiations that helped the company improve its terms. It spotted trends in maintenance requests that helped them prevent problems before they happened.

One of my favorite examples is a healthcare clinic that completely transformed its patient experience. They were facing typical healthcare challenges - long wait times for appointments, high no-show rates, and overwhelmed staff trying to manage everything.

We implemented a comprehensive AIgent system that handled scheduling, reminders, follow-ups, and basic patient inquiries. The AIgents could understand medical terminology, recognize urgent situations, and seamlessly coordinate with the human team when needed.

The results were dramatic. No-show rates dropped by 60%. Patient satisfaction scores increased by 45%. But here's what really impressed me - the staff reported feeling less stressed and more satisfied with their work. They could focus on patient care instead of administrative tasks.

Let me share a manufacturing success story that demonstrates how AIgents can transform operations. This company was struggling with inventory management and production scheduling. They had frequent stockouts of some items while others gathered dust on shelves. Production schedules were always playing catch-up with demand.

Their AIgent system didn't just manage inventory - it revolutionized their entire production process. It analyzed historical data, seasonal patterns, and market trends to predict demand. It coordinated with suppliers, managed production schedules, and maintained optimal inventory levels.

Within six months, they reduced inventory costs by 28% while improving product availability. Production efficiency increased by 35%. But the most interesting outcome? They discovered new market opportunities because the AIgents spotted demand patterns they hadn't noticed before.

Here's a case that shows how AIgents can help small businesses compete with larger competitors. A local auto repair shop was losing customers to big chains that could offer 24/7 service scheduling and quick response times.

We set up AIgents to handle their customer communications and scheduling. These AIgents could discuss car problems with customers, schedule appointments, send reminders, and even provide basic maintenance advice. They integrated with the shop's diagnostic systems to provide status updates on repairs.

The shop owner told me something interesting after six months: "We're not just competing with the big chains now - we're beating them." Their customer retention rate went up by 40%, and they were getting new customers specifically because of their responsive service.

Let me share one more case that really shows the power of AIgents in transforming business operations. A financial advisory firm was struggling to scale its client services. Their advisors were spending too much time on routine client inquiries and basic portfolio reviews, limiting their ability to take on new clients.

We implemented AIgents that could handle routine client questions, provide basic portfolio information, and even generate preliminary investment recommendations based on client profiles. The AIgents could recognize when questions needed human expertise and seamlessly transfer those conversations to advisors.

The impact was significant. Advisors increased their client capacity by 45% while actually providing more personalized service. Why? Because they could focus their time on complex planning and relationship building instead of routine tasks.

These success stories share some common themes. In each case, AIgents didn't just automate tasks - they transformed how the business operated. They freed humans to focus on work that really needed their skills. They improved consistency and quality of service. And they uncovered opportunities for improvement that nobody had noticed before.

In the next chapter, we'll look at how you can implement AIgents in your organization. You'll learn the step-by-step process we use to achieve results like these.

8

Implementing AIgents in Your Organization - A Practical Guide

Let's get down to the nuts and bolts of bringing AIgents into your business. I've shown you what's possible - now I'm going to show you exactly how to make it happen. This isn't theoretical stuff. These are the exact steps we use with our clients to ensure successful implementation.

First, let's talk about assessment. Before you implement any AIgents, you need to know where they'll make the biggest impact. I always start by asking clients three key questions:

What tasks are eating up your team's time? Where are the bottlenecks in your operations? Which areas could deliver better results with faster response times?

Let me give you a real example. A law firm came to us wanting to implement AIgents across their entire operation immediately. Ambitious, but not the best approach. When we did our assessment, we found that document review was consuming 40% of their associates' time. That became our starting point. By focusing on one high-impact area first, we could show clear results and build momentum for wider implementation.

Next, you need clear goals and metrics. Vague objectives like "improve efficiency" aren't enough. You need specific, measurable targets. When we worked with that law firm, we set clear goals: reduce document review time by 50%, maintain 99% accuracy, and free up at least 15 hours per week per associate for higher-value work.

Now let's talk about choosing the right AIgents for your needs. This is where many businesses go wrong. They try to force generic AI solutions into their specific workflows. That's like trying to fit square pegs into round holes.

Take customer service AIgents. A retail business needs AIgents that can handle product inquiries and order status updates. A healthcare provider needs AIgents that understand medical terminology and privacy requirements.

A financial services firm needs AIgents that can manage sensitive financial information and compliance requirements. Same basic technology, but very different implementations.

Here's something crucial - training your team. I've seen great implementations fail because organizations didn't properly prepare their people for working with AIgents. This isn't just about technical training. It's about helping your team understand how AIgents will make their jobs better.

Let me tell you about a manufacturing company that got this exactly right. Before implementing AIgents in their quality control process, they involved their QC team in the planning. They showed them how AIgents would handle routine inspections, allowing them to focus on complex quality issues. The team went from skeptical to enthusiastic because they could see how it would improve their work.

Integration with existing systems is another critical factor. AIgents need to work seamlessly with your current technology stack. When we helped a healthcare provider implement AIgents, we had to ensure they could interact with their electronic health records system, scheduling software, and billing systems. More importantly, we had to ensure all this integration maintained HIPAA compliance.

Here's my recommended implementation sequence:

Start with a pilot program in one department or process. This lets you prove the concept and work out any issues before rolling it out more widely. A regional bank we worked with started with loan application processing in one branch. Once they saw it working well, they expanded to other branches and processes.

Monitor everything closely during the pilot. You're not just looking for efficiency gains - you're looking for ways to optimize the AIgent's performance and identify any necessary adjustments. When we implemented AIgents for a retail chain's inventory management, the pilot showed us that we needed to adjust how the AIgents prioritized restock alerts based on actual sales patterns rather than just inventory levels.

Get feedback from everyone involved - both the team members working with the AIgents and the customers or end-users affected by them. This feedback is gold. It helps you refine the implementation and identify opportunities for improvement you might have missed.

One thing I always emphasize - be prepared for some initial resistance. It's natural. People get nervous about change, especially when it involves AI. The key is to show, not tell. Let your team see the AIgents in action. Let them experience how it makes their work better.

A customer service team I worked with was initially worried that AIgents would make their interactions with customers feel robotic. But when they saw how the AIgents could handle routine inquiries instantly, freeing them to spend more time with customers who needed human attention, they became enthusiastic supporters.

Here's something else that's crucial - ongoing support and maintenance. AIgents aren't "set it and forget it" solutions. They need regular monitoring and updates. But they also get better over time as they learn from more interactions.

The good news? This maintenance is much easier than managing traditional software. AIgents can identify their own performance issues and suggest improvements. They can adapt to new situations without requiring complete reprogramming.

Let me give you a specific example of how this works. We helped a real estate company implement AIgents for property management. Initially, the AIgents handled basic tenant inquiries and maintenance requests. But over time, they learned to predict common maintenance issues based on patterns they observed. They started suggesting preventive maintenance schedules that reduced emergency repairs by 40%.

Security and privacy considerations are also crucial during implementation. You need clear protocols for data handling, access controls, and compliance requirements. This is especially important in industries with strict regulatory requirements like healthcare, finance, and legal services.

One approach I've found effective is to implement AIgents in layers. Start with handling public or less sensitive information, prove the security measures work, then gradually expand to more sensitive areas. A financial services firm we worked with started with AIgents handling basic product inquiries before moving to account-specific information.

Finally, let's talk about measuring success. You need both quantitative and qualitative metrics. Track the obvious numbers - time saved, tasks processed, response times. But also look at indirect benefits - employee satisfaction, customer feedback, new opportunities identified.

Remember that law firm I mentioned earlier? Beyond the obvious metrics of faster document review and cost savings, they found that their associates were providing better service to clients because they could spend more time on strategic legal work. That's the kind of comprehensive success you're looking for.

In the next chapter, we'll dive into the ethical considerations of implementing AIgents. Because doing this right isn't just about technology - it's about creating positive change for your business and your people.

9

The Ethics of AIgents
- Doing AI Right

Let's talk about something that matters a lot to me - implementing AI the right way. Ethics isn't just a buzzword when it comes to AIgents. It's about making sure this powerful technology helps your business while respecting your employees, customers, and values.

I'll be straight with you - AI raises important questions. How do we protect privacy? How do we ensure fairness? How do we maintain transparency? These aren't just theoretical issues. They affect real people and real businesses every day.

Let me tell you about a situation that came up with a healthcare provider we worked with. They wanted their AIgents to handle patient scheduling and basic health inquiries. Great goal, but we had to think carefully about privacy. Medical information is sensitive - you can't treat it like regular customer service data.

We designed their AIgents with privacy built into every interaction. They verify identity before sharing any information. They know exactly what they can and can't discuss in different situations. They maintain detailed logs of all data access. Most importantly, they're programmed to err on the side of caution - if there's any doubt about privacy, they transfer the interaction to a human.

Here's another ethical challenge we deal with often - fairness in decision-making. AIgents can process vast amounts of data and make quick decisions, but we have to ensure those decisions don't discriminate unfairly. A financial services company we worked with wanted AIgents to help screen loan applications. We had to be extremely careful to ensure the AIgents weren't inadvertently biasing decisions based on factors like zip codes or names.

The solution? We built in regular bias testing. The AIgents flag any patterns that might indicate unintended bias. Human managers review these patterns and adjust the systems as needed. It's not enough to just say we're being fair - we have to actively work to ensure fairness.

Transparency is another crucial issue. People have a right to know when they're interacting with an AIgent instead of a human. But there's more to transparency than just disclosure. It's about making sure people understand what AIgents can and can't do.

One of our retail clients handles this beautifully. Their AIgents clearly identify themselves at the start of each interaction. They explain what they can help with and how to reach a human if needed. When they make recommendations, they can explain the reasoning behind their suggestions. This builds trust with customers because there's no attempt to deceive.

Let's talk about data security. Every business must protect its data, but different businesses have different security needs. A law firm handling sensitive case information needs different security measures than a retail store managing inventory. We work with each client to implement appropriate security protocols based on their specific situation.

One thing I always emphasize - you need clear policies about what data AIgents can access and how they can use it. A manufacturing company we work with has AIgents that analyze production data to optimize processes. But these AIgents don't have access to employee personal information or proprietary design data. Clear boundaries matter.

Employee concerns are another ethical consideration we take seriously. People worry about AI affecting their jobs. The key is to be honest and clear about how AIgents will be used. Show your team how AIgents will help them, not replace them.

A customer service department we worked with handled this well. They involved their team in planning how AIgents would be implemented. They showed exactly which tasks AIgents would handle and how this would free up humans for more interesting work. Result? The team became enthusiastic supporters because they saw how it would make their jobs better.

Here's something else to consider - AIgents need to align with your company's values. If your business prides itself on personal service, your AIgents should enhance that, not detract from it. A high-end retail client of ours uses AIgents to provide better personal service by handling routine tasks so their human staff can focus on building relationships with customers.

Accountability is crucial too. When AIgents make decisions or take actions, someone needs to be responsible for those outcomes. We help businesses set up clear oversight processes. Who monitors AIgent performance? Who reviews decisions? Who handles exceptions? These aren't just operational questions - they're ethical ones.

Let me share an example of how this works in practice. A real estate company uses AIgents to screen rental applications. The AIgents can process applications quickly, but all rejections are reviewed by human managers. This ensures efficiency while maintaining fairness and accountability.

Then there's the question of continuous improvement. AIgents learn from their interactions, but we need to ensure they're learning the right things. Regular audits of AIgent behavior help catch any developing issues early. Are they maintaining consistent service quality? Are they treating all customers fairly? Are they protecting privacy appropriately?

A financial services firm we work with reviews their AIgents' interactions weekly. They look for any patterns that might indicate bias, check privacy compliance, and ensure the AIgents are making appropriate decisions about when to escalate to humans.

Here's something I want to emphasize - ethical AI isn't just about avoiding problems. It's about creating positive change. When implemented properly, AIgents can help make businesses more fair, more transparent, and more responsive to human needs.

Think about customer service. AIgents can provide consistent service to everyone, regardless of the time of day or how busy things are. They don't play favorites or have bad days. When designed with ethics in mind, they can actually help reduce bias and improve fairness.

The bottom line is this - implementing AIgents ethically isn't just the right thing to do, it's good business. Customers trust companies that handle their data responsibly. Employees engage better when they understand and trust the technology they're working with. Regulators look more favorably on businesses that take ethics seriously.

In the next chapter, we'll look at future trends in AI technology. But remember - as AI capabilities grow, our commitment to ethical implementation becomes even more important.

10

Where AIgents Are Heading – The Future of Work

et me tell you something exciting - we're just scratching the surface of what AIgents can do. Every month I see new capabilities that amaze me. But more importantly, I see new ways these capabilities can transform how businesses operate.

I was talking with a client recently about how AIgents have evolved just in the past year. When we first implemented their customer service AIgents, they could handle basic inquiries and route complex issues to humans. Now those same AIgents can understand context, emotion, and nuance in ways that seemed impossible a year ago. They can have natural conversations that flow just like human interactions.

But that's just the beginning. Let me show you where this technology is heading and what it means for your business.

First, AIgents are getting dramatically better at understanding complex situations. Think about a typical business problem - it usually involves multiple factors, some of them conflicting. Traditional software struggles with this kind of complexity. But new AIgents can analyze dozens of variables simultaneously and make nuanced recommendations.

A manufacturing client of ours is testing AIgents that can balance production schedules, inventory levels, supply chain constraints, energy costs, and maintenance needs - all in real-time. They don't just follow rules; they understand trade-offs and can adjust strategies as conditions change.

Voice technology is another area seeing massive improvements. The AIgents we're developing now don't just understand words - they understand tone, emotion, and cultural context. They can adjust their communication style to match the person they're talking to. This means more natural, more effective interactions.

One of our healthcare clients is implementing voice AIgents that can understand not just what patients say, but how they say it. If someone sounds stressed or confused, the AIgent adjusts its approach accordingly. It's the kind of emotional intelligence that used to be purely human territory.

Here's something really exciting - AIgents are getting better at working together in coordinated teams. Imagine having different AIgents handling various aspects of your business, but all communicating and coordinating with each other seamlessly. One AIgent spots a trend in customer behavior, another adjusts inventory predictions, while a third modifies marketing strategies - all automatically.

We're seeing this already with a retail client. Their AIgents work as a coordinated team - customer service AIgents share insights with inventory AIgents, which communicate with supply chain AIgents, creating a smooth, integrated operation that would be impossible to manage manually.

Predictive capabilities are advancing rapidly too. The next generation of AIgents won't just react to situations - they'll anticipate them. They'll spot patterns humans might miss and identify potential problems before they occur.

A financial services firm we work with is testing AIgents that can predict customer needs based on subtle patterns in their behavior. If a customer's transaction patterns suggest they might be planning a major purchase, the AIgents can proactively provide relevant information and support.

But here's what really excites me about the future - the growing ability of AIgents to enhance human creativity and decision-making. They're becoming true partners in the creative process, offering insights and suggestions while leaving the final decisions to humans.

I saw this recently with a marketing team. Their AIgents don't just handle routine tasks - they analyze market trends, customer behavior, and campaign performance to suggest creative new approaches. The human team still drives the creative process, but they have insights and possibilities they wouldn't have discovered on their own.

Data analysis is reaching new levels too. Future AIgents won't just crunch numbers - they'll find meaningful patterns and explain them in ways that help humans make better decisions. They'll turn raw data into actionable insights automatically.

A real estate client is working with AIgents that don't just analyze market data - they identify emerging neighborhood trends, predict property value changes, and suggest optimal timing for transactions. It's like having a team of expert analysts working 24/7.

Security and privacy capabilities are advancing alongside these other improvements. New AIgents can adapt their security protocols based on the sensitivity of the data they're handling. They can detect potential security issues more effectively and respond more quickly to threats.

But let me be clear about something - the future isn't about AIgents replacing humans. It's about creating more effective partnerships between human intelligence and artificial intelligence. Every advance in AI technology opens new opportunities for human creativity and strategic thinking.

Think about how your smartphone has changed how you work. It hasn't replaced your brain - it's enhanced your capabilities. Future AIgents will do the same thing on a much larger scale. They'll handle routine tasks more effectively while giving humans better tools for creative and strategic work.

The businesses that thrive in this future will be those that find the right balance between human and artificial intelligence. They'll use AIgents to handle routine tasks and process data, freeing humans to focus on innovation, relationship building, and complex problem-solving.

We're already helping clients prepare for this future. It's not about implementing every new AI feature as soon as it appears. It's about understanding which advances can truly benefit your business and implementing them in ways that enhance rather than disrupt your operations.

The key is to stay flexible and keep learning. The AI landscape will continue to evolve rapidly. But if you understand the fundamentals and work with experienced partners, you can take advantage of new capabilities as they emerge while maintaining stability in your operations.

Remember - the goal isn't to chase every new AI trend. The goal is to build a more efficient, more capable organization that combines the best of human and artificial intelligence. That's the future we're helping our clients create, and it's more exciting than anything I've seen in my career.

Are you ready to be part of this future? In our final chapters, we'll look at exactly how AIgent, LLC can help you get there.

11

Why AIgent, LLC Should Be Your AI Partner

L et me be direct - there are a lot of companies out there offering AI solutions. So why should you choose AIgent, LLC? I want to show you exactly what makes us different and why it matters for your business.

First, let's talk about expertise. As a certified MindStudio AI Expert, I've helped countless businesses implement AI successfully. But more importantly, I understand that AI is just a tool. What matters is how that tool solves real business problems and creates real value.

I remember talking with a potential client who'd already spent a fortune on AI solutions that didn't deliver. *"Everyone promises the moon,"* he told me. *"But nobody seems to understand what my business actually needs."* That's exactly why I started AIgent, LLC. We don't sell technology - we solve problems.

Let me show you how this works in practice. When a new client comes to us, we don't start by talking about AI capabilities. We start by understanding their business. What are their challenges? What's eating up their team's time? Where are the bottlenecks in their operations? Only then do we start designing solutions.

Our partnership with MindStudio is crucial here. They're the leaders in AI technology, and our partnership gives us access to capabilities other companies simply can't match. But it's not just about having powerful tools - it's about knowing how to use them effectively.

Think about it like this: having a professional-grade kitchen doesn't make someone a great chef. What matters is knowing how to use those tools to create something valuable. That's what we do with MindStudio's technology. We combine cutting-edge AI capabilities with deep business understanding to create solutions that actually work.

Here's something else that sets us apart - we're obsessed with results. I can't tell you how many times I've heard horror stories about AI implementations that looked great in demos but fell apart in the real world. That's not how we operate. We measure success by your success.

Let me give you a concrete example. A manufacturing company came to us after a failed AI implementation with another vendor. They'd spent months and a lot of money on a system that their team wouldn't use because it didn't fit their actual workflows. Within three weeks of working with us, we had AIgents operating smoothly in their processes, and their team was actually excited about using them.

Why the difference? Because we took the time to understand how their team really worked. We designed AIgents that fit their processes instead of forcing them to change everything to fit the technology. And we stayed with them, making adjustments and improvements based on real-world feedback.

That brings me to another crucial point - we're in this for the long haul. We don't just implement AIgents and walk away. We stay with you, ensuring the technology keeps delivering value as your business evolves. Think of us as your AI partner, not just your AI vendor.

This ongoing partnership matters because AI technology is constantly evolving. New capabilities emerge regularly. But you shouldn't have to worry about keeping up with every new development. That's our job. We monitor the technology landscape, identify improvements that could benefit your business, and help you implement them effectively.

Security and reliability are also top priorities for us. We understand that AIgents handle critical business functions. You need to know they'll work consistently and securely. That's why we implement multiple layers of protection and monitoring. We help you establish appropriate security protocols based on your specific needs and regulatory requirements.

But here's what really makes us different - we understand that successful AI implementation isn't just about technology. It's about change management. It's about helping your team embrace new ways of working. It's about creating positive change in your organization.

That's why we put so much emphasis on training and support. We don't just show your team how to use AIgents - we help them understand how AIgents will make their jobs better. We work with you to communicate effectively with your team, address concerns, and build enthusiasm for the changes.

Results matter. A healthcare provider we worked with saw their patient satisfaction scores increase by 45% after implementing our AIgents. A manufacturing company reduced their operational costs by 28% while improving product quality. A law firm freed up 40% of their associates' time for higher-value work.

But you know what I'm most proud of? The feedback we get from employees who work with our AIgents. Things like "This actually makes my job better" and "I can finally focus on the work I was trained to do." That's when you know you're doing AI right - when it enhances human work instead of just replacing it.

Let's talk about something practical - cost. AI implementation is an investment, and you need to know you'll see returns. We work with you to identify and track specific metrics that matter to your business. Whether it's time saved, costs reduced, or revenue increased, we help you measure the real impact of AIgents on your bottom line.

Looking ahead, partnering with AIgent, LLC means you'll be ready for the future of AI. As new capabilities emerge, you'll have a trusted partner to help you evaluate and implement them effectively. We'll help you stay ahead of the curve without chasing every new trend.

Think about where your business could be six months from now. Imagine your team freed from routine tasks, able to focus on strategic work. Picture your operations running more smoothly, your customer service improved, your costs reduced. That's what we help our clients achieve every day.

Ready to learn more? In the next chapter, we'll look at specific next steps for getting started with AIgents. I'll show you exactly how we can help transform your business with AI done right.

12

Taking Action – Your Next Steps with AIgents

et's get practical. You've seen what AIgents can do. You understand how they can transform your business. Now it's time to talk about making it happen. I'm going to lay out exactly how to get started with AIgents in your organization.

First, let me share something I've learned from implementing AIgents in hundreds of businesses: success isn't about moving fast - it's about moving smart. The businesses that get the best results aren't the ones that try to transform everything overnight. They're the ones that take a strategic, measured approach to implementation.

Here's what that looks like in practice. Start by identifying one area of your business where AIgents could make an immediate impact. Maybe it's customer service, where your team is overwhelmed with routine inquiries. Maybe it's document processing, where hours are wasted on manual data entry. Maybe it's inventory management, where you're struggling to maintain optimal stock levels.

I worked with a law firm recently that did this perfectly. Instead of trying to automate everything at once, they focused on document review - their biggest time sink. Within weeks, their AIgents were handling initial document reviews, freeing up associates for more valuable work. The success in this one area-built momentum for wider implementation.

Next, gather your data. What are your current metrics in this area? How much time does your team spend on these tasks? What are your error rates? What's it costing you? You need this baseline data to measure the impact of your AIgents.

But don't just look at numbers. Talk to your team. What frustrates them about their current work? What tasks do they wish they could spend less time on? What would help them be more effective? This input is invaluable for designing AIgents that truly enhance your operations.

Now, let's talk about your first implementation. At AIgent, LLC, we use a proven process to ensure success:

Assessment: We work with you to understand your specific needs, challenges, and goals. This isn't a generic evaluation - it's a deep dive into your business.

Design: We create AIgents specifically for your operation, integrated with your existing systems and workflows. This is where our MindStudio partnership really shows its value.

Testing: We run pilot programs to ensure everything works smoothly before full implementation. This lets us catch and address any issues early.

Training: We work with your team to ensure they understand how to work effectively with their new AIgent colleagues. This isn't just technical training - it's about helping them see how AIgents make their jobs better.

Implementation: We roll out your AIgents systematically, monitoring performance and making adjustments as needed.

Ongoing Support: We stay with you, ensuring your AIgents continue to deliver value and adapt as your needs change.

Let me tell you about a manufacturing company that followed this process. They started with inventory management - a constant headache for them. During the assessment phase, we discovered they were spending hours manually updating stock levels and generating purchase orders.

We designed AIgents that could monitor inventory levels, predict demand patterns, and automate ordering processes. But we didn't stop there. During testing, we found opportunities to integrate with their supplier systems, making the entire supply chain more efficient.

The training phase was crucial. We showed their team how AIgents would eliminate the tedious parts of their jobs, letting them focus on supplier relationships and strategic planning. By the time we implemented fully, the team was enthusiastic about working with their new AI colleagues.

That's the kind of transformation we can create in your business. But it starts with taking action. Here are your next steps:

Choose your starting point. What area of your business needs transformation most urgently?

Gather your current metrics. You need this baseline to measure success.

Talk to your team. Their input is crucial for successful implementation.

Contact us for an initial consultation. We'll show you exactly how AIgents can transform your specific operation.

Remember - the businesses that thrive in the coming years will be those that effectively combine human and artificial intelligence. Every day you wait is a day your competitors might be getting ahead.

Let me be clear about something: this isn't about replacing your team. It's about empowering them. It's about eliminating the routine tasks that drain their time and energy. It's about letting them focus on work that really matters.

Think about where your business could be six months from now. Imagine your operations running more efficiently. Picture your team focusing on strategic work instead of routine tasks. Envision your customers getting faster, better service.

That's not just possible - it's happening right now in businesses that have implemented AIgents effectively. And we can make it happen in your business too.

Ready to take the next step? Contact us. Let's talk about transforming your business with AIgents. The future of work is here - and we're ready to help you make the most of it.

13

Realizing the Full Potential of AIgents - Your Path Forward

Let me leave you with something important. Throughout this book, we've looked at what AIgents can do, how they work, and how to implement them. But there's a bigger picture I want you to see - we're at the beginning of a fundamental shift in how business works.

Think about how mobile phones transformed business. At first, they were just for making calls on the go. Now they're essential tools that have changed how we work, communicate, and run our operations. AIgents are following a similar path, but the impact will be even more profound.

I was talking with a business owner recently who said something that stuck with me. Six months after implementing AIgents, he told me, "This isn't just about doing things faster or cheaper. We're doing things we couldn't even attempt before."

That's the real power of AIgents. They don't just improve existing processes - they enable entirely new possibilities. Let me show you what this looks like in practice.

A financial services firm we work with started using AIgents for basic customer service. But as they saw what was possible, they began expanding their vision. Now their AIgents proactively identify customer needs, spot potential problems before they occur, and help design personalized financial solutions. They're not just serving customers better - they're reinventing what customer service means.

Here's what I want you to understand: The companies that thrive in the coming years won't be the ones with the biggest AI budgets or the most advanced technology. They'll be the ones that best combine human and artificial intelligence. They'll be the ones that use AIgents to enhance what their people can do, not replace them.

I see this playing out in every industry. Manufacturing companies using AIgents not just for automation, but for predictive maintenance and adaptive production planning. Healthcare providers using AIgents not just for scheduling, but for coordinating complex patient care. Law firms using AIgents not just for document review, but for identifying patterns and insights in vast amounts of legal data.

But here's the key - success with AIgents isn't automatic. It requires clear vision, strategic implementation, and the right partner. That's why we built AIgent, LLC the way we did. We're not just a technology provider. We're a partner in transformation.

Let me be direct about something: If you're reading this book, you're already ahead of many of your competitors. You understand that AI isn't just coming - it's here. You see the potential for transformation. But potential only matters if you act on it.

Think about your business right now. What tasks are eating up your team's time? What opportunities are you missing because you're caught up in routine work? What could you achieve if you could multiply your team's capabilities?

These aren't theoretical questions. They're the starting point for real transformation. Every business I work with starts with these same questions. The ones that succeed are those that move from questions to action.

Remember that law firm I mentioned earlier? They asked themselves these questions and took action. Now their associates spend 40% less time on document review and 40% more time on strategic legal work. Their clients get better service, their team is more engaged, and their business is growing faster than ever.

The manufacturing company that implemented AIgents for inventory management? They're not just saving money on inventory costs. They're identifying market trends faster, adapting to changes more quickly, and outperforming their competitors.

The healthcare provider using AIgents for patient scheduling? They're not just filling appointments more efficiently. They're providing better patient care because their staff can focus on people instead of paperwork.

These transformations are possible in your business too. But they require taking that first step. They require moving from interest to action.

As we wrap up this book, I want to leave you with three key thoughts:

First, the AI revolution isn't coming - it's happening now. The businesses that wait to adapt risk falling behind.

Second, successful AI implementation isn't about replacing humans - it's about enhancing human capabilities. AIgents should make jobs better, not eliminate them.

Third, you don't have to figure this out alone. That's why AIgent, LLC exists. We're here to help you transform your business with AI done right.

The future of work is being written right now. Some businesses will lead this transformation. Others will follow. Some will thrive. Others will struggle to catch up.

Which path will your business take?

The choice is yours, but you don't have to make it alone. We're here to help you succeed with AIgents. Ready to take the next step? Let's talk about transforming your business.

92

The end of this book isn't really an ending - it's a beginning. It's your invitation to join the businesses that are already transforming how they work with AIgents. The future is waiting. Let's build it together.

About the Author

Tony Eggleston is an accomplished entrepreneur, AI expert, and Amazon best-selling author of *"How YOU Can Use ChatGPT."* As the founder of AIgent, LLC, Tony has dedicated his career to helping businesses harness the  power of artificial intelligence to drive productivity, improve customer experiences, and create innovative, future-ready workplaces. Certified as a MindStudio AI Expert—the highest level of expertise available through MindStudio—Tony brings deep knowledge and hands-on experience in deploying AI solutions that blend seamlessly with human workflows.

Tony's work is driven by a commitment to making AI accessible and beneficial for all. Through AIgent, LLC, he has guided numerous organizations in various industries to achieve their operational goals by implementing AIgents, intelligent digital assistants designed to streamline workflows, enhance customer service, and empower employees.

With a passion for educating others about AI, Tony writes and speaks extensively about the transformative impact of AI in today's workplace. His insights make complex technology approachable, and his expertise provides readers with practical strategies to navigate the evolving digital landscape. In *AIgent: AI Agents and the Future of the Workplace,* Tony sh ares his vision for a collaborative future where AIgents and human employees work side by side to create more efficient, innovative, and productive organizations.